Treat Climate Change, Save the Earth:

How to Prevent Flooding and Drought to Slow Global Warming

Amanda Rothman

Table of Contents

DEDICATION

To my Father who we recently lost, and my Mother who is still with me.

To my family and friends, and Humanity sharing our planet.

To the SPS gang, and all the people who made this project a reality!

Introduction- Why Read This Book?

First of all, let me make it clear. This book is not a debate on whether climate change exists. This is a call to action!

Our world leaders are searching for a way to **slow the causes** of climate change.

I am writing this book to give you information about another method we can use to bring many problems under our control. An infrastructure we can build which can **treat the symptoms** of flooding and drought!

- Have you ever had to suffer through thirst, heat waves, dying crops, or water restrictions due to a drought?

- Have you ever had to slog through flooded streets, wondering how you were going to deal with the damage to your house and property?

- Or worse- have you ever been frantic with worry about your family and friends after a hurricane or wildfire destroyed their homes?

- Do you feel sad when hearing about severe weather which lead to thousands of tragic deaths in other countries?

- Do you feel frustrated that so many people agree that something needs to be done, but few have any idea of where to start?

The discussion about whether humans have caused climate change has already been written and debated. What I'm doing for you is outlining a realistic way to take action and solve some of our largest problems.

We Need The Right Tools

There are so many people out there already doing what they can to make a difference. But their voices and ideas are only being heard by hundreds or thousands of people. It sounds like a lot, but it's not enough to get worldwide attention.

Good ideas are being developed, but without funding and support, they aren't used as they deserve to be.

Entrepreneurs are developing innovative ways to:

- Filter and desalinate water

- Create personal energy generators which use water, solar, and wind

- Create new methods of dousing fires without harm to the environment

I have an idea of my own. An idea which could use many of the methods mentioned above as part of its structure. An idea which can solve many different problems on a global scale if it is taken up by world governments.

But What Can I Do, You Ask?

As I mentioned earlier, in order to gain government attention, ideas involving an undertaking this large would also need to be supported by the public.

An idea that "goes viral" in a short amount of time often gains attention from world leaders. Wouldn't it be nice for something viral to be a source of potential good news instead of something tragic, or something fun but fleeting? Information that we can use to begin to find more solutions for world problems that are slowly escalating?

What you can do is to share news about this book! Social media is of course very helpful. I will maintain a list of my accounts where I post regularly on my blog. You could join me there too; I'll keep you up to date on any progress!

www.proactive-action.com

If you know any politicians or people in news outlets, you'd be handing them something that I'm hoping will be very helpful to them!

The solutions that I'm sharing with you in this book are those related to flood and drought.

These two symptoms are intertwined in their causes and effects. Flooding and drought can both be treated at the same time using the system I have designed.

We can work toward a common purpose whether or not we "believe" in climate change.

The proactive system I have outlined in this book includes a means to:

- Safeguard our water's cleanliness, including accidents or acts of sabotage

- End drought in stricken areas so our farms can flourish

- Protect communities from flooding due to powerful storms

- Bring wildfires under control more easily

- Create enough hydroelectric energy to run this system and perhaps even to power our homes, reducing our dependence on fossil fuels and even nuclear power

- Work to rebuild and restore our glaciers to prevent sea-level rise and help to lower the temperature of the oceans

Even though this proactive system will require repairs to our current infrastructure as well as new additions, this work must be done with attention paid to preserving our natural resources. Construction must be done in a way that is thoughtfully and strategically planned out.

The last thing I would want is for untouched areas to be disrupted for the sake of this project when one of the main concepts behind this project is to help nature restore itself!

The Current State of Events

Before I get into the system itself, we need to start on the same page so that we know where to go from here:

In recent years, there has been a noticeable increase in temperature and changes in weather patterns, recently put under the neat umbrella called "climate change". Some well known examples:

- We are encountering a new phenomenon called "Super Storms". These storms are characterized by their larger-than-normal size and record-breaking barometric pressure despite being the remnants of earlier hurricanes. Famous examples are Super Storm Sandy in 2012 which battered the New England coast, and Super Storm Nuri in 2014 which hit Alaska and only caused less damage due to a sparser population in the area.

- Hurricanes and Typhoons are growing stronger than ever. A prime example is Hurricane Patricia, the strongest storm yet on record which developed to Category 5 in a matter of a day or two over very warm ocean waters. And astoundingly, there were two major hurricanes which reached the Arabian Sea: Tropical Cyclones Chapala and Megh!

- Extreme weather causes widespread power outages due to lightning strikes or the weight of snow on the power line structures. This was also experienced during Super Storm Sandy and other winter time storms in the northern parts of the country.

- Drought like the ones in the Midwest and California spanning the past several years threatens our food supplies, cripples our agricultural industry, creates difficult living conditions for the poor living there, and strains the global economy.

- According to the US Geological Survey reports, the drought in California is also causing the valley land to sink. When people keep pumping the groundwater so vital to their survival, the land above it sinks steadily over time. These lowered areas can damage infrastructures like roads, bridges, aqueducts, and flood control. The instabilities are believed to be linked to earthquakes in the area.

- Dry areas across the world become susceptible to wildfires which can be caused not only by lightning, negligence, or arson, but also from the

12

sheer level of heat and UV rays due to our depleted ozone layer.

- Wildfires threaten lives, property, and the livelihoods of many. We risk our brave emergency personnel, and spend valuable resources fighting blazes that affect thousands of acres at a time. The air we breathe grows thick with ash and the chemicals used to fight against nature. These fires give off massive amounts of carbon.

- At the same time, sometimes only a few thousand miles away, people experience floods of near biblical proportions. For example, in 2012 the US Midwest down through Texas was suffering a crippling drought. Earlier this year the same areas were flooded from a series of damaging storms.

- More recently was the South Carolina flood of 2015 which burst through a number of dams and threatened the integrity of many others. The rivers swelled and waters rose, killing people, sweeping away and destroying property, and eroding the stability of the land we depend upon.

How This Project Developed

I've been studying weather trends my entire life. It's been a hobby and a passion of mine. I've also done extensive experimentation on what I like to call "world building" with groups of intelligent and creative people. Running theoretical situations like these helped me to envision this system and plan for many safeguards.

Over the past years, I'd been keeping track of the increasing trend of the worst disasters we've ever seen, causing lost lives, billions of dollars of damage, and people left to live without power or potable water for weeks.

I heard other reports detailing droughts that last for years like the one in California, record temperatures in the Middle East killing people without shelter, and wildfire after wildfire causing death, damage, and displaced families.

- I thought how terrible it was that there were people in one place desperate for water and people in the same country who were drowning due to raging rivers.

- Then there were reports about burst water pipes and the waste of millions of gallons of now tainted drinking water across the US, even in drought-ridden areas like California.

- There were reports about chemical toxins leaking into the water supply of Charleston, West Virginia, and the surrounding counties in 2014, requiring that residents live on bottled water.

- There were reports about accidental oil spills like the famous Deepwater Horizon spill in the Gulf of Mexico, and burst oil pipelines like the one off of the California coast polluting our seas and beaches, killing wildlife, and destroying the way of life for hundreds of thousands of people.

- There were reports about ships unable to travel along the Mississippi River due to record low levels of water during the drought of 2012.

- There were reports about aquatic wildlife migrating out of the tropics from areas where they had spawned for centuries because of the increases in sea water temperatures. Some aquatic species such as the ones living in the Gulf of Mexico have nowhere north to migrate to cooler waters unless they somehow manage to find their way to the Gulf Stream, so they are endangered.

- There were reports of icebergs melting, destroying Arctic habitats, and adding water to the oceans worldwide.
  --At the North Pole, sea-based glaciers melt, leaving fewer surfaces above water available for polar bears and other arctic animals to live on. This threatens the continuation of those species. The glacier melt also affects the water temperature and UV/ozone levels, which in turn generates more melted ice.

- There were reports about sea levels rising around the world, threatening coastal US states like Florida and many areas across the world which are not far above sea level.

--At the South Pole, Alaska, and Greenland among other places, **the land-based glaciers melt and add the entire amount of glacier water to the total water level of the oceans.** This affects sea-level rise even more drastically than glaciers from the North Pole, many of which are already floating in water.

Report after report after report of senseless tragedy.

I'd had enough!

Is there a solution to these problems? The proactive system I present in this book can treat the symptoms of climate change, quickly relieve the man-made problems, and even solve some of our energy shortages in the process.

But it will require an investment in our future. An investment from everyone. From individuals like you who are reading this book to decision-makers in both the corporate and government sectors, both domestically and around the world.

A proactive investment for a proactive system. Right now!

What Is Being Done Now?

The current efforts toward staving off climate change are very different from what my system would accomplish. But they are vitally important!

The Environmental Protection Agency (EPA) is already partnering with local and international groups to further the research and conservation efforts into advancing climate change science. This global awareness will be an enormous help with our future efforts to come together to advance such good causes.

There are so many fantastic projects and groups taking part in the series of Climate Change Summits being held in addition to the UN's session in Paris in 2015. The main event is the Conference of the Parties to the United Nations Framework Convention on Climate Change (COP21/CMP11), otherwise known as Paris 2015. There have already been a number of festivals springing up around France to show off additional projects and to share public enthusiasm.

President Barack Obama and other world leaders have agreed to meet to discuss what can be done to lower the levels of carbon usage. This will be the 21st session, and a crucial one which will outlast the changing of the leaders in any country.

I am more than confident that our world leaders will achieve a new international agreement. This agreement is intended to be applicable to all countries, with the aim of keeping global warming below 2°C. These first steps toward global cooperation are the most important ones; their success will make worldwide treatment of flood and drought possible.

I was also thrilled to hear about Pope Francis' Encyclical, Laudato Si, in which he laments global warming and the degradation of the environment. I was even more thrilled that the Vatican released versions in different languages so that I was able to read it! This treatise on the need to make global changes for the sake of humanity and the care of our planet is well written and thoughtful.

Pope Francis has many important suggestions and hopes for our world, and in his recent tour has not been shy about speaking about them. I highly recommend taking a look at the Encyclical, released here on the Vatican's website and translated into English. He emphasizes the need to work with the Earth rather than against it, to respect the planet instead of merely living on its surface and cluttering it with our worldly possessions.

I'll reemphasize how this proactive system I'm suggesting must include this requirement: That we build around and preserve nature rather than plowing through it. The entire idea behind this proactive system is to allow the Earth to naturally rebuild its freshwater sources while allowing us to more easily survive using other resources.

Symptoms vs. Causes of Climate Change

The efforts of our leaders to control carbon use is an entirely separate issue from what I will discuss. Their plan is to work to control one of the main CAUSES of these weather changes. Currently, scientists are working on reducing the level of carbon emissions across the planet, and to continue to increase awareness of the subject. I support and applaud their work! I hope that you will too.

But now that they have begun the difficult task of hunting down the cause, I believe it is necessary to triage and treat the SYMPTOMS.

These symptoms are the consequences which humanity is suffering. Flooding and drought. Super storms. An increase in ocean temperature which is causing marine life to migrate and glaciers to melt.

Problems like these symptoms need to be treated at the same time as the causes. We have to come at these problems from an entirely different direction so that we're not relying on one long-term method to take care of everything.

In this book, I'll be discussing a proactive system that we can use.

What is this System?

The system that this book describes in detail is a method for transporting water.

It has similarities to our current infrastructures for transporting natural gas, and for transporting water on the **micro** scale. A simplified description of the micro scale we already use involves transporting water from water treatment plants to individual houses, and waste from individual houses and storm drains back to water treatment plants.

The system that I am hoping to help build involves water transport on the **macro** scale, transporting large quantities of water from water tanks and bodies of water to where it is needed, even if it is across a country or across a continent.

This macro transport system will involve a number of important parts such as a way to filter water more thoroughly while it is being transported

- It will begin with a water filtration system far improved than our current technology.

- It will involve a pump system to keep the water moving as directed.

- It will move water through pipes across the country from areas which are threatened by flood, and deliver it to drought-stricken areas, or into storage tanks for later use.

- As water moves through the pipes, it will be further filtered. The act of transporting the water through generators will produce hydroelectric energy.

- In order to carefully monitor the water and know where it is needed, another part of the system will be used as the "brain."

I'll go into detail about each part of the system in the rest of this book.

Wouldn't it be a wonderful irony if we were able to stop using the Pipelines for transporting oil and replaced them with ones to help transport water instead?

What are Other Advantages and Benefits?

A solution with such far-ranging effects as this one inevitably depends on a number of working parts. But with such a huge potential for immediate and future benefits, I believe the investment would be more than worth it:

- A large number of jobs will be created in order to first build and then maintain this proactive system, boosting our global economy.

- There are many research opportunities out there, available to individuals and to corporations if they are forward-thinking enough to invest in them. Not only are they a potential great source of public relations, but the resulting technology would be profitable as well.

- It can bring about a decrease in disease and the necessity of emergency humanitarian measures. It could perhaps even soothe the violence that comes from the human nature of being justifiably angry and frustrated due to the difficulty of survival in some areas of the world.

- Without drought and with better crop harvests and livestock, food prices would go down.

- With fewer people getting sick and having better access to healthier foods, it could lower medical bills.

- It could prompt companies to lower insurance rates in the face of fewer disasters.

- With our natural freshwater resources being restored, there would be a probable reduction in earthquakes and wildfires.

- With a new source of energy, it could eliminate the need for fracking and reduce the need to transport coal and oil. Hopefully this would also mean fewer accidents on trains, and fewer burst oil pipelines and rigs.

It might be obvious from these ideas that I'm hoping this proactive system will be shared worldwide.

A New Way of Solving World Problems

As we mature as humans, we also need to improve the way that we develop solutions to solve our problems.

A solution **cannot** bring worse problems to the table.

Ideally, any solution should be able to support or pay for itself, or even offer multiple benefits. If there are problems, the benefits must outweigh them. The problems must be solvable. With this proactive system in place, that will be within our reach.

My particular suggestions for this water transport system will solve a number of the issues that I've listed. It will also benefit job growth and increase our potential for creating green power sources, reducing pollution, and generating the actual energy required to run the proactive system. It's possible that done properly, it can generate enough excess energy to provide power to our planet's populated areas!

Done with the necessary research, we can carry out this proactive system with a minimal effect on the environment.

This is not a system that can be constructed by the lowest bidder. This proactive system is a major investment in our planet's future, potentially for centuries.

My greatest wish is to protect and improve the lives of people on Earth, but I do not want to do so if the cost is our natural resources. This is particularly important to me as I adore nature.

Why Now? Why Not Later?

Unfortunately, the task of controlling the causes of climate change will take many years to carry out, and even longer for the effects to be felt. Especially because the hard work of our planet's leaders will require agreement, followed by careful compliance that lasts throughout further elections and changes in regime.

In the meantime, the world is suffering from a number of symptoms which also need attention. Every year that passes only means more damage done, more lives senselessly lost.

When these tragic symptoms are reduced or even halted, we'll have more time and resources to bring the causes themselves to a standstill.

But in order to bring about these treatments in the near future, we need to begin acting NOW.

But Really-- What Can I Do About It?

Who can help to bring this about? You!

This is because you are the one reading this book. You care enough to have read up to this point.

I BELIEVE IN YOU.

Don't think that your silent agreement will be enough (although I'm glad it's there). The word needs to get out. Information must be spread to people in influence.

Even if my ideas are just discussed with friends and family, it will be of help. But the true way to get this proactive system - or any system - created is to make those making the decisions aware of these ideas!

I will be doing my utmost to bring this to the attention of people in power, but hearing from one lonely soul won't even make this idea a blip on anyone's radar.

With your help reading this book and spreading the word, these ideas will hopefully find their way into the minds and hearts of those with the authority and the resources. They can begin to investigate and carry them out in a manner beneficial to us all, and to our future!

If you are one of those people, I urge you to consider these ideas. Not only will they benefit you, your company, or government, but also the planet, and humanity itself.

Conclusion

If we look at the world as our body, there are a number of biological systems needed for it to function properly. Water resources are involved in a number of these biological systems, like the circulatory system, the nerves and brain, and the digestive and excretory systems.

In this chapter, I gave you some of the background, motivation, and the ideal future of this proactive system, and its benefits to us all.

In the next chapter, I'll go into water filtration, its vital role, and the many ways it can be used to improve the lives of the entire population of Earth, beyond providing clean water.

Chapter 1 - Water Filtration

When we discuss the vital systems of our bodies, the digestive and excretory systems are among the most vital, but they are certainly not the first ones we might mention in polite company. But I'm afraid that I must do so here in this book.

By taking in and filtering water from various sources like rivers prone to flooding, we are doing the equivalent of eating and digesting the nutrients that we require in order to live. The materials that we filter out of the water are removed by excreting them, but we will take this a step further. By using green methods, we want to take the waste materials and re-purpose them into something useful!

Water filtration is the first phase of this new system. We would take in water from non-freshwater sources, and make them potable for human use.

Water Filtration and How It Can Save the World

Scientific research is a vitally important role in the construction of this proactive system. Water filtration is one of the most important tasks. We filter our water now, but with the proper level of water filtration, we could take water from anywhere in the country and use it in any other area.

Currently, the sources of water that we use are a fraction of the water available on the planet. These rare and precious sources of fresh water are mostly located within the confines of our landmasses, whether they are underground or bodies of water hemmed in and flowing over the continent's crust.

As our population and our need for water grows, the strain on these sources of freshwater will increase until those resources are sucked dry. On top of our own needs, the land and wildlife need a certain amount of water in order to support themselves in natural areas. A depletion of this water is one of the reasons behind drought and wildfires.

The best action we can take is to stop relying on these natural sources of water and let them replenish themselves. Instead, we could build a method to filter water from the ocean, and use that water to support our population.

The Benefits of Improving Water Filtration Technology

Let me begin by quoting a report from the EPA about water and its current use in energy production:

> 'The use of water is critical to many aspects of energy production, including extraction of energy resources (e.g., mining), refining petroleum, transporting fuel by barge along waterways, and generating electricity through hydropower or thermoelectric power (where water is used as a coolant). In addition, water resource management itself uses a lot of energy (e.g., supply, distribution, and treatment of water and wastewater consume about 4% of US power generation (USDOE, 2006)). Due to symmetry in the use of water to produce energy and the use of energy to produce (and deliver) clean water and treat wastewater, increased efficiency of either the production or use of energy or water can yield many benefits throughout the economy.'

Research beyond what we have already invented must begin now if we are to develop a new technology to work with the proactive system in this book as soon as possible. Currently our methods of filtration work, but they aren't as efficient as we need them to be. There are even some new and greener methods currently in development, but we need to be able to adopt them properly if they turn out to be the solution that we're looking for.

Fortunately, our computer technology has advanced in ways that were not available to us a mere decade ago. The power and capabilities of supercomputers continue to grow to an extent where agencies like NASA are able to use fluid dynamics and other methods to test different configurations of airplane wings before building any physical models. These same supercomputers could be used to help research new water filters and many other aspects of this proactive system.

Not only do we need a new filtration technology for our own use in this system, but also for the world's benefit. There are so many places lacking in clean water where people fall victim to disease because of that lack.

Even the upcoming 2016 Summer Olympics will be affected as the water in Brazil's capital city of Rio de Janeiro has been reported as unsuitable even to boat in without high risk of getting sick from the level of viruses present in the water.

Arid areas could be made to flourish with new sources of water. I wouldn't exactly recommend filling in the Grand Canyon and making it into additional farm land, but there are regions across the world that can be given the opportunity to become self-sustaining with clean water as an easily accessed resource.

Further on in this book, I will outline ways we can potentially use water in this proactive system to create a near continual source of green energy.

The Current State of Filtration

We currently have many ways of filtering our water. These methods need to be improved and standardized. Water tested in some areas still shows some levels of pollutants.

Despite that, some basic research into the water and waste management plants will show the astounding level of work that goes into water treatment as it stands now

The EPA has already begun public awareness about the quality of local drinking water. An impressive amount of information can be found online.

If we are to transport and use water across the entire country, the existing filtration technology must become much more efficient and effective. Portability of the filtration system would be a big bonus. We'd be able to transport and use the filters in humanitarian efforts as well!

We will be filtering and transporting an enormous amount of water every year, hopefully for centuries to come. Our current infrastructure has endured for hundreds of years in some areas. What we build here and now must be built to last!

I'll be describing research and investment opportunities for this proactive system further in this chapter and throughout the book, but water filtration is a very large and very important one! These opportunities aren't my own; I'm just trying to bring ideas into being.

The Water is Different Across the World

Different methods of water filtration result in different qualities of water with different impurities remaining. These results are also related to the quality of water at its source. Some natural water springs in one area might be high in a mineral content that can't be completely filtered out using the methods used at that local water and waste management plant.

Other sources of freshwater across the country might be located in a swampy area. There might be a lot of different types of wildlife using it, and salt water from the ocean might occasionally seep in to the freshwater reserves. The method used at this local water and waste management plant may be completely different, yet still unable to completely cleanse the water of those influences.

The water is safe to drink by the standards we've devised, but it would not be a good idea to deliver water from the swampy area in order to irrigate the area containing natural springs. The content of the water from one area can unbalance the ecology in the other area, like transporting wild animals to live outside their native habitats.

We might not want to think about it, but there are natural microorganisms living in our water, and a chemical content native to the local area. This is why if we are to build a proactive system where we can use water anywhere in the country, it must be cleansed at a much higher level!

The Ideal Future of Water Filtration

Some ideal qualities of a new filtration technology would make the filters easy to clean and reuse, or have some part of them biodegradable and able to be easily changed.

These filters would need to clean water to the degree that there are no remaining particles in the liquid- just water - H_2O. This way the water can be used anywhere without upsetting the ecology if any gets out into natural water sources. This would also make it easier to spot any impurities, which I will describe later.

Research is a vital industry, but there is only so much money and a limited number of researchers who can be put into it. I believe that it would do more in the long run to assign these resources to research water filtration.

The ideal method should be standardized for ocean water. This would mean including a desalinization step, plus additional methods of cleaning or separating ocean water polluted by oil, human waste, and other chemicals.

A bonus would be a portable filtration method which could be sent out into oil slicks in the ocean to clean water. The ideal method must not be a risk to ocean life. Researching and producing this method for cleaning out oil can be at least partially funded by oil companies as a prudent investment in the future to enhance their public images. It could even be a profitable technology for them.

Thoughts on Corporate and Private Contributions

The oil companies or other large corporations can help create and adopt this technology as a new method for keeping the companies profitable in the face of competition from an increasing number of renewable sources of energy instead of relying on profits from drilling for oil. The filtration technology can become a gateway to move oil companies into focusing more on green energy.

This would be a smart move, because if my system and the rules for reducing the causes of climate change were to be successfully adopted, then there would be a significant reduction in the need to rely upon burning fossil fuels. Between that and already low oil prices, it's a no-brainer.

Additionally, instead of research resources going to companies creating new and horribly expensive drugs to cure diseases from drinking dirty water, we could prevent those diseases by providing this system to create properly purified water.

Again, as with the case of oil companies taking part in this technology, the same can be done with the willing aid from pharmaceutical corporations. If these large corporate entities invest into this proactive system, they can profit from it while helping to fuel the efforts to improve life for everyone.

Universities and laboratories that are already enthusiastic in their involvement with climate change might be another wonderful source of assistance. Information and knowledge are critical resources, and the use of supercomputers is a great way to gain that information. Use of existing supercomputer resources from universities and laboratories, plus any other donations of equipment would get us to our ideal advancements more efficiently.

There are already some amazing private entrepreneurs who have developed new methods for cleaning water, and generating and storing energy. They just need more time and resources to make these methods easier to produce. With people like these, this system can come into being that much sooner!

Getting to the Ideal Filtration From Where We Are Now

The ideal would be a new technology which can handle all impurities, but this may take a while to create. In the meantime, we need to build the system NOW and it needs to have water filtration as its base.

So, for the moment, we should:

1. Build the water filtration system using the best of the current filtration technology that we have available

2. Build the rest of the system that I'm detailing in this book to be flexible enough to easily modify when new filtration technology is developed

3. Invent the new filtration technology

It is possible to have many different layouts of filters. We already have a number of these different systems as a part of our waste management plants. Here is what I would suggest we do with filter placement in the new proactive system:

- They can be placed at the sources and destinations of the water.

- They can be integrated as part of the transport and storage systems.

- They can be concentrated at the water management sites.

But no matter the arrangement, the filters need to be easily repairable or replaceable while maintaining a high level of security.

The best option that I see for us to begin with is to make the filtration a multiple stage filter in different areas of the system.

Stage One: Physical filters at the water intake spots

These filters shouldn't permit plant, animal, or large formations of minerals to enter the system. The physical filters can have several layers, getting smaller with each one until we have only uncleaned water in the pipes. This would also be a good spot for the desalination step. Desalination is the process of making saline water into freshwater.

Physical filters must be built and placed with special care. Some areas of water intake would include the ocean. Coral reefs are a vital part of our shorelines.

They must be protected for the sake of the marine life, not to mention our own safety and livelihoods.

As part of these water intake areas, it would be wise to also add some features to shelter the reefs from damage that they're already suffering. This could include shielding them from the increasing levels of UV and damaging storms, as well as to keep them from being accidentally damaged from this proactive system - or even from sabotage.

As for where the water intake units would be, that might be deeper into the ocean where they would have less of an effect on the coastal area.

Stage Two: Chemical filtration

A series of filters can treat water along transportation pipes and remove unwanted chemicals. The water will be pushed through several of these filters as it moves through the transport part of the system. There would be a careful check of the water at the entrances of populated areas. If chemical toxins are detected either through accident or sabotage, the system would shunt the water away to be treated. I will go further into this in the next chapter.

Stage Three: Filtration at the water processing plants

I call these Water Houses and will detail them in a later chapter. It is quite likely that this level of water cleansing will require a lot of energy, as it does today. With the methods for generating energy that I will detail in later chapters, these processes can be run without using energy from our current electrical grid. In fact, it is entirely possible that there will be enough energy generated through this proactive system to power emergency services to communities, or even whole populated areas.

There are several good methods of water cleansing which can be used until a future ideal can be invented. I mentioned general processes above, but some specific ones are: reverse osmosis, water synthesis (which is another avenue for research to make it an energy efficient method), and Biogas Cogeneration (CHP).

If you're interested, here is more information on the <u>possible use of Biogas Cogeneration on wastewater</u>

Handling Waste From Water Filtration

Material removed from water must be repurposed in some way. This would be in keeping with the ideal philosophy of creating a solution that doesn't cause additional problems. For example:

Physical material like plants and leaves, minerals, and the occasional fish can be made into compost that can be either sold/given to farmers.

Chemical toxins can be analyzed and treated. Ideally, this would neutralize the toxin or create a different and useful material. This process can be fueled by this system if necessary. With future research, it could also be fueled by the energy given off by chemical reactions.

Non-toxic waste can be treated chemically and neutralized, but in a different location than toxic waste. There's no need to mingle these together unless it has been conclusively proven that there are no better uses for any non-toxic materials.

Water: More Than Just for Human Use!

Water filtration is vital if we are going to be able to tap our oceans as a water source and let our natural freshwater reservoirs replenish themselves.

The rising sea level is a threat. One way we can avoid that is to use our seas as a water resource.

If our water filters are good enough, we can re-purpose that water for more than just human use. I'll explain further uses as we go on deeper into this book.

Conclusion

In this chapter, I explained about our current state of water filtration and ways that it needs to be improved in order to create this new proactive system of ours. These filters, the digestive and excretory systems of our body, will be closely connected to the circulatory system of our "body".

The circulatory system equivalent is the largest and most vital part which will transport, filter, and even generate energy to run the system of our body: The water pipes infrastructure!

Chapter 2 - Pipes Infrastructure

Here is where we really get into the meat and potatoes of this system. The pipes infrastructure is the circulatory system of the 'body' of our country. Just as the circulatory system transports blood to deliver food, oxygen, white blood cells, and other vital fluids, and transports wastes to be removed from the body, this system of pipes also has many functions.

Blood is directed where it needs to be, taking in waste and delivering oxygen. The water in pipelines will be similarly directed. In order for this to work, the pipelines will have a number of items built into them at periodic intervals. These will include filters, toxin traps, valve-like structures to direct water into different pipes, hydroelectric generators, and sensors used to detect water pressure and water quality.

Recent Events Point Toward Our Future Ideal

As we've heard on the news, there have been a number of major breaks in the water or sewage systems. Incidents have been reported in areas like the drought-stricken California. A 90-year-old water main deluged Sunset Boulevard and the UCLA campus, wasting hundreds of thousands of gallons of water in less than an hour. The flood also created a 15-foot sinkhole.

Broken water pipes waste millions of gallons of water that would have been drinkable. Areas get flooded, causing property damage and affecting our economy. Water also can become tainted when water pipes burst, prompting communities to have to boil water or purchase bottled water to use from day to day until the repairs are completed and water is tested and safe to use again.

We already have methods for water storage and filtration, but those must be expanded. Transporting water by truck from place to place the way that we transport many of our liquid goods like milk and gas would be an unreasonable solution except perhaps in extreme emergency.

Instead, we will need to follow the example of what we've used for so long.

We'll need to repair, replace, and expand our current water pipes infrastructure. Much of it is more than a century old.

Transporting water at such a large scale throughout the entire country will mean a large investment of time, money, materials, and effort up front. But if this is done correctly and with the proper materials, we can make the system last through the next generations while allowing it to be easily maintained.

This sort of system can solve several of our major problems in one blow. We can replace areas of water pipes that have been weakened by centuries of use, strengthen our infrastructure, improve the security of our resources, and avert the threat of our seas rising. In doing so, we can begin to cure the major symptoms of climate change.

Water Transportation

The main purpose of the pipes infrastructure is, unsurprisingly, shunting water from place to place. I use the word 'shunt' particularly because the act of moving the water to different locations is a part of what will run this system. The path of the water will be directed by other parts of the system detailed later, but the results will be:

1. We can move water from levels of high concentration to levels of low concentration

- Proactive weather management- When storms are predicted to cause flooding, water in that area can be taken into the system before the storm hits, and shunted away so that the rain event will replenish the water that has been removed.

- If the storm is stronger than predicted, more water can be removed as needed. If the storm is weaker, water can be brought back to the area from storage and pipes.

2. We can deliver water as needed to areas that need it

- For human use - we need potable water for drinking, cleaning, and the other obvious day-to-day necessities and luxuries.

- Irrigating crops - we can bring an end to droughts killing our crops and also keep our livestock healthy. Stronger plants and animals will be more resistant to diseases, which have also been at an increase in recent years.

- Tank storage - we can store water for shortages and emergencies. For example, if toxins managed to get into the system, that portion of the water within the pipe infrastructure can be cut off and treated. In the meantime, there is an uninterrupted supply of water for use.

- Fighting wildfires - we can swiftly and continuously deliver water to our emergency crews so that fires can be doused more quickly and with less damage, and more importantly, less risk to their well-being.

- Increasing the water retention capabilities of soil - we can research the best methods to deliver water into soil that is suffering from drought so that future storms won't result in too much water run-off leading to flooding. One such method is soil aeration, which I will discuss in a later chapter.

- Managing the system's water pressure - we can prevent burst pipes and other issues by having control over the amount of water flowing through them and the direction the water travels in.

3. We can filter the water and generate energy for the system and more by transporting the water through the system.

Transporting Water on the Macro Scale

We already collect water from filtered and processed sources, and deliver the water to individual homes and commercial buildings. This could be considered a **micro** transportation system.

What we need is to create a **macro** transportation system. This is a method to move the water in larger quantities from locations where there is plenty, and deliver it to depleted areas or water storage tanks rather than individual homes.

This system will transport water between filtered water sources from one part of the country to another, using pipes in the same way that oil is transported. However, where many people protest the transport of oil through their property, it's likely that there would be far less reluctance to allow water to pass through.

The places that the pipes are built on should be charted so that there would be as little disruption as possible to the natural surroundings. As I have mentioned earlier, distance is not a factor. In fact, it is an advantage to have the pipes take a longer route in order to avoid natural areas. The more distance water travels in a pipe, the more it will be filtered, and the more energy it will potentially generate. We just need to be strategic about where the pipes are located.

Improving Water Transport on the Micro Scale

We have a system of water filtration and delivery from water plants to individual locations, but this system is old and in some areas is also breaking down. The crumbling infrastructure results in burst water pipes and the loss of hundreds of billions of gallons of cleaned water every year. We need to replace this system with similarly solid materials that we use for the transportation of water on the macro scale.

There have been many news reports detailing burst water pipes flooding roads and buildings. We've also heard about water sources being tainted by areas of chemical storage or oil leaks. The filtration system can go a long way toward making sure that water remains potable, and to quickly take care of the problem if it is contaminated. It could even deliver a stored excess of water to areas that are in need of it while the local problems arising are taken care of.

Pipes Materials

The pipes themselves need to be made from a material that won't degrade over the years whether they are underground or not, and that won't allow chemicals of any sort into the water. Additionally, the pipes should be easy to clean and non-reactive to chemicals used for cleaning or other purposes.

Like any system of hoses, a water pipe infrastructure has the greatest risk of breakages at the weakest points. That is why we can't completely rely on our current system which is already experiencing occasional pipe bursts. Water pressure monitoring will help prevent this, but it is best to be proactive about building a system that is to last for centuries. Especially with a growing world population which will create a parallel increase in need for water.

It would be best if the material has a certain amount of flexibility in case of an earthquake or other event, yet at the same time it needs to be strong enough to withstand pressure or punctures from the outside. One possibility for profitable research and invention is a micro-woven material like some modern car exteriors. Filaments for the purpose of transmitting the location of any percussive or disruptive impact could be added to the pipes in a way that would be transmitted to sensors and couldn't be disabled.

Lastly, as one might expect, the pipes would need to have some method for changing the path that water takes, much like switches on train tracks alter the path of a train. If water is needed in one area, and needs to be moved away from another area, that path needs to be made available. If water needs to be sent to storage towers in a certain location, those commands must be sent and automatically carried out. The method for determining this will be discussed in the next chapter on Water Houses.

Currently, we use PVC pipes in new water mains and sewage construction, but iron needs to be used in areas where the water pressure is particularly high. These new pipes will work for a while, but eventually they too will be affected by time and wear. We need to research better materials for our pipes that will last centuries into the future.

Temperature Control

Another ideal quality for a new pipe material would allow for temperature control. In extremely low temperatures, water in pipes can freeze and slow or stop water travel. In high temperatures, water can evaporate and cause a build in air pressure which must be averted so there is no chance of pipes bursting.

Pipe materials with temperature control can be used in drainage areas so that an enormous build-up of snow can be slowly and safely melted. Therefore a sudden increase in air temperature like we experienced in the Winter of 2014-2015 doesn't threaten a flood to a snowed-in town. Once water pipes, sewer pipes, and storm drains are replaced, snow melted and cleared from the roads can be diverted away.

This feature could be added in several ways. They could be a device dotted along the pipe infrastructure like water filters would be. Or the pipe material could be the sort that is able to drain away heat or cold, 'neutralizing' the temperature. Alternatively, there could be some sort of material that can be activated to melt snow as needed without heating it up. Experimentation with thermodynamics is in order!

As with my previous ideas, an ideal technology would allow us to use these temperature extremes and generate energy. At the very least, the energy taken from hot water could be used to raise the temperature of cold water in another area.

Perhaps in the future we could find a way to create energy from cold water. For now, we could always repurpose the cold water to a local Water House to help keep its computer systems or the building itself cool.

Water Tanks and Storage

We already have and regularly use large water tanks in populated areas. We need more of them that are able to be easily sanitized and built to filter and maintain the purity of water stored for extended periods.

Empty water tanks also will be used as needed to assist with keeping water pressure stable in the system. If the water pressure is too high in a particular area of the pipe's infrastructure, water can be drained into these waiting tanks so that there will be little risk of future burst pipes.

Small water tanks outside towns can be used in conjunction with toxin traps or as emergency pressure water storage.

Water Filtration

The water pipes' infrastructure is the second area of water filtration, mostly involving treating water chemically. The first area of filtration is at the pipe's intake areas for sea water and emergency canal/river water.

Filters run throughout the pipes so that river/canal water can be taken into the system at any point necessary to prevent flooding. This also means that water will be filtered/treated multiple times, possibly using different methods, in order to optimize its purity.

Pipes outside towns may contain water not yet completely filtered. Pipes inside towns contain drinkable water only. Water processing plants can be located there as well as within their current buildings in order to cover all eventualities.

Toxin Traps

Directly outside of populated areas is a special system of pipes used as toxin traps. This is where water that has been tainted by accident or sabotage is delivered for emergency treatment so it doesn't flow inside the towns.

There also can be toxin traps located in the water management buildings to guard against drinking water being tainted. If necessary, tainted water located inside a town can be purged from the system through a fire hydrant and delivered to a toxin trap by truck.

Toxin traps function by filtering water repeatedly in a loop of pipes built with several different filters. This part of the system can be closed off from the main system. Water will not be released back into the system until it is tested negative for pollutants.

These pipes also will have energy generators installed in them at periodic intervals, using the water that is being transported through the pipes to generate energy. This will power the system detoxifying water as well as the other parts of the system, making it self-reliant.

Sensors and Security

Of course, a water system for the future will need continual management and an extremely accurate system of real-time information delivery; this is the nervous system of the body. This will mean a secure system of sensors.

With our current level of technology, we can manage this even now, but as with water filters, flexibility must be allowed in order to later install better technology years later. Research should go into features like sending real-time data to the water management buildings, including analysis of water content in the area as well as the water pressure. This would make a huge increase in the security of the system.

Water analysis sensors that can periodically test water content would have many other profitable uses for the company that invented it, meaning more investment opportunities for the research.

Sensors will report on water pressure in the pipes, where the water is moving to, the content of the water in the pipes, the necessity of cleaning/replacing filters, and notifications of where tampering or accidents have been located.

This information from the sensors will be managed by the water management buildings, or Water Houses discussed in the next chapter.

Energy generation

Here is one of the crown jewels of this entire solution. The system I'm describing in this book can be self-powered. Through the act of filtering and transporting water, there is an enormous potential to create energy through hydroelectric generators in the water treatment plants as well as along the entire pipe infrastructure.

The energy created by these generators can be used to power the water filters, water pumps, the sensors, and the water management buildings. In this way, the system does not need to rely on external sources of energy. It won't drain our resources, and it will be less vulnerable to sabotage.

If these new hydroelectric generators create enough energy, they can even power emergency services such as hospitals, police, and fire stations for populated areas, and possibly even be the power sources for homes of the population of towns and cities.

This is the ideal that we would work toward. The system will pay for itself far beyond our current methods of water delivery. I'll be going more into energy generation in the next chapter.

Conclusion

I just went over the circulatory system of our body, the pipes infrastructure. However, these 'blood vessels' will be far more functional than our own. They will be closely integrated with the Nervous System, sending real-time information to our brain. This brain is what I will explain about in the next chapter: Water Houses.

Chapter 3 - Water Houses

In this chapter I'll be outlining Water Houses, and their part in this system that we would build.

In the way that the pipes' infrastructure is the circulatory system, the Water House and the sensors are the brain and nervous system. It will collect data from many sources from the system and more, such as detailed global weather analysis in order to accurately predict where water needs to be moved.

The complex nature of the system means that human minds will need to be the force behind the decisions made based on the information collected by the Water Houses. So whenever I mention the Water Houses, I also mean the people running them.

It's possible a new branch of science/degree/security clearance might result from the creation of this new system. People with security clearance would be best to work in the Water Houses. They would need to have a good familiarity with weather phenomena as well as a background in chemistry, geology, and computer systems.

What are Water Houses?

Water Houses are structures similar to existing water management buildings. However, their specialized and additional functionality means it would be easiest to build new structures rather than completely rely on the existing water management buildings.

The system we're building can definitely use existing buildings to increase the overall efficiency of water distribution. However, there will need to be many more of these buildings than there are in our current water management system.

Water Houses can be located inside of towns to assist with localized water management, but their main purpose is to manage the water in the country-wide system of pipes. Since transporting water over such long distances through the pipe infrastructure requires constant water pressure, the Water Houses will also act like a computer signal booster to keep water continuously moving on the path where it is supposed to go.

The Water Houses also would be built with protection against fire and other natural disasters. Because of the sensitive nature of protecting our water resources, Water Houses also work together to maintain security. If a Water House somehow sustains damage, the other Water Houses can be used in its stead once it has been confirmed that repairs are needed. A high level of security would be needed to keep the buildings safe and protected from sabotage/tampering.

The computer system contained in the Water Houses would need to especially secure. Because of the vital functions they perform, they CANNOT be allowed to be hacked or disaster can happen.

The Final Level Of Water Filtration

As I mentioned in the first chapter, there are several good methods of water cleansing that already exist which can be used until a future ideal can be invented. Some examples are: Reverse osmosis, water synthesis using the water as a material (research), Biogas Cogeneration (CHP) used on waste water among them.

Not only can we use supercomputers to help invent and test new filtration technologies and strategically plot out the pipes infrastructure, but we can also use them to build and test out which methods of water filtration would be best to use as the last step in the Water Houses.

It is entirely possible that different methods would work better for different areas of the country due to the different types of water contaminants that are local to the area. This should also be kept in mind when the Water Houses are strategically placed.

There are pollutants that seep into lakes, rivers, and even seawater which are common to most of the country, but it is likely that that won't be the case with all countries across the world. Also, accidental pollutants in areas that have been extensively mined will be different from the accidental pollutants in areas that are built over swamps. Water Houses will be able to track issues like that and be flexible enough to make the necessary changes.

Critical Information Analysis

Water Houses, being the brain (or in this case, multiple brains) of our body, would receive the information sent from the other areas of the body. Sensors will send information from a large number of places:

The water pressure in the pipe infrastructure:

The water pressure in the pipe infrastructure needs to be maintained at a certain level so that there is no risk of pipes bursting, and so that water travels at a consistent rate throughout the system.

The water levels in each water tank across the country:

As I discussed in the chapter on pipe infrastructure, different types and sizes of tanks will be used for different functions. They can store drinking water, stabilize the water pressure inside the pipes by removing or adding water as needed, and remove tainted water from the system to be treated safely.

If a populated area is being threatened by flood, large water tanks can be used to increase the speed that water is removed from rivers and lakes.

Of course, the Water Houses will track when the water was taken in, and what the content of that water is so that it can be used properly later, either filtering it into the pipes' infrastructure or returning it to the rivers and lakes when they are low enough.

The tested contents of the water throughout the system:

Water must be tested regularly. Ideally, all of the water will be tested as it flows through the pipes once the sensors are powerful enough to do so. This information will be monitored by the Water Houses in real time so that necessary changes can be made as needed.

For now, random periodic testing is the more likely option. Agencies like the US Geological Survey's Water Watch could be a good place to start for how to do this real-time testing properly. When a better sensor technology has been developed, the flexibility of the water pipes structure would allow us to install the new sensor array.

The functional status of the system:

This includes the physical condition of other Water Houses, toxin traps, filtration system, the entire pipe infrastructure, existing water management buildings, the hydroelectric generators, the power plants, the computer systems, and any other parts of the system we build into it.

The Water Houses receive notifications of any sort of tampering, natural damage, or wear and tear that requires maintenance. This could include anything from changing or cleaning filters, tanks, and pipes to unblocking a water intake.

The real-time weather information like the National Hurricane Center and other resources provide:

Water Houses would use this weather information in order to take proactive action on where water needs to be moved to and from. Ideally they could also get information from their own radar/weather setups as a backup system.

The measurements of soil dryness vs. water retention, and the locations that need water from other areas:

Water Houses would receive test data on soil testing over large areas. If soil is too packed and dry, it can't absorb water as easily. This is one of the reasons why drought-stricken areas get so much water runoff when you'd expect that the soil would absorb the moisture like a sponge.

Water Houses can put out action requests to aerate the soil in an area, to plant trees and grasses native to the area, etc. There will be more on this in the next chapter on Proactive Weather-Based Water Management.

If Water Houses are built well with forward-thinking flexibility in mind, more functionality can be built into them as we find new ways to make this system more efficient!

Keeping The Water In Motion

The Water Houses need to contain a pump system which works in concert with other Water Houses across the country. I described Water Houses as a signal booster in a computer system. This is because without this sort of pump, water would slow down in long stretches of the pipe infrastructure.

The insides of pipes aren't frictionless, and water has to be pushed through the filters and hydroelectric generators. High water pressure can't be used for this without risking damaging the pipes in that area. So we add pumps to the system. Possibly before and after the filters and generators.

This is a "living" system, meaning it can and will develop and grow into something new depending on what we need from it. We're making it flexible just for this particular reason.

Proactive Weather-Based Water Management

This not-quite new field of work has the responsibility of deciding when and how to move water away from areas where storms are threatening a flood. This water would be moved into tank storage or into the water pipes' infrastructure depending on the water pressure in the pipes system. By doing this, water is basically being moved from areas of forecast high concentration and toward areas of lower concentration.

We already have a system like this in some states, and it works well. But it can be improved upon.

Water Houses would be used to monitor and direct water flow when we are aware of strong storms approaching an area. Humans and computers analyze the information about storm forecasts, ocean level rise from surges, etc. Using this information and the information about the current state of water in the pipes, tanks and reservoirs, the Water Houses plot out paths of water, working with other Water Houses for confirmation.

A powerful computer system would be able to track the current levels of all affected areas, including natural water reservoirs as well as man-made storage water tanks. It would also assist in calculating what actions are best to take when a new water shift out of an area is necessary.

Tracking the water is somewhat similar to tracking storms, only with fewer random variables because the pipe infrastructure is a closed system. The Earth is a closed system as well, but it's much larger and the weather system is enormously complex.

Updating Our Weather Analysis Methods

The necessary weather analysis is a continuous process, just like the Earth's weather is continuous and ever changing. Everything that happens in Earth's weather affects the weather somewhere else, like a line of dominoes. Or, in this case, like a globe of dominoes where the dominoes can reset themselves.

Weather isn't a domino, though. Winds shift direction and strength. They contain water, or dry air. Pressure goes up or down. Sometimes an electric charge builds and creates lightning. Temperature changes. Under the right conditions, spirals are formed into hurricanes or tornadoes.

It's time to start tracking weather globally rather than tracking and analyzing it country-wide. We already do this to an extent, but not widely. We have the computer technology to be able to analyze everything that happens on our planet. We just need to decide to use computers that way. We would know with certainty if a cluster of storms is going to develop into a hurricane or not.

Plotting Out the Strategic Path for the Water

The weather analysis I described above would be a large source of the information used to make decisions on how to handle the water in our system.

Using the pipe infrastructure, we can lower the current level of water in a flood-threatened area in preparation of an approaching powerful storm. We can drain excess water during and after the storm to prevent it from reaching damaging levels. This water can be filtered and delivered to depleted reservoirs in other areas of the country and storage tanks near and far.

The water temperature also would be taken into account, moving it to best effect for cooling hot water and repurposing water that is getting too cold.

While water could be moved and used country-wide with the right water filtration, until we reach that level of water cleanliness, we should begin by keeping water use more localized.

Currently, water contains material from local areas. This includes microorganisms and minerals from the local ecosystem. Until these things can be removed from the water, it would be best to use the water locally. This way, we don't risk ecosystems being harmed by non-native microorganisms or minerals it doesn't need, and allows the ecosystem to receive those things it depends on.

The information from soil and water analysis will enable the Water Houses to plot out the area that would best be able to use that water on the micro scale.

Using Water Houses on the Micro Scale

The Water Houses direct water through the pipe infrastructure, water filters, toxin traps, water storage tanks, water management buildings, and hydroelectric generators. There are also water distribution channels like fire hydrants, homes, buildings, sprinklers, and the sewer system. These distribution channels are examples of water management on the micro scale instead of the macro scale.

Fire hydrants are especially important for emergencies. We must ensure the water is there when it is needed, and with enough water pressure to be useful.

But there could be alternate uses for the fire hydrants, adding even more functionality to an existing system. For instance, if water has been tested and deemed toxic inside of towns, the nearest fire hydrants can be used to drain that water out of the system and into trucks with water tanks. The toxic water can then delivered to toxin traps and cleaned.

Or, if our water filtration technology has reached an ideal level, we could have toxin traps/water filters dotted throughout a populated area to guard against tainted water. Towns wouldn't have to boil water or rely on bottled water again.

Sending Out Other Action Requests

Using the information sent to the Water House "brains", they can send requests for maintenance until these tasks become more automated. A few examples from the system as I've described it so far:

- Manual requests to move water to or away from an area

- Changing or cleaning water filters

- Processing and repurposing filtered waste

- Using and cleaning toxin traps

- Using and cleaning water tanks

- Sending a truck to drain tainted water to be cleaned

- Putting in requests for an area to undergo soil aeration

Energy generation

Like the pipe infrastructure I described in the previous chapter, Water Houses can have their own hydroelectric generators, as well as an independent system of pipes going around the house itself. Energy generated this way would be independent of the rest of the system.

It could be used as security to make the Water Houses self-sufficient, and if enough energy is generated, to help power the local population. This will reduce our dependency on the forms of fuel that contaminate our environment.

These hydroelectric power generators work with existing power plants to deliver energy to it in the same way individuals with solar panels can sell their extra energy to the power plants. Only this system is meant to work together, so ideally the energy wouldn't be sold.

In other words, the system that delivers our water will also deliver at least some of our power needs. This power can be generated by the transport of water through the pipes and Water Houses, and it can even be generated by shunting water through a smaller localized system of pipes, filters, and generators that is built around the Water House itself.

Scientific research is a never-ending process, and this is the case for this system as well. Water Houses can be upgraded with new technology. More efficient water filtration systems can be developed and installed. Better energy transformers can be created, increasing the amount of energy safely transferred without loss to entropy.

If we can become more reliant on this and other green forms of energy, we can move away from nuclear energy as well. This would mean that energy production could no longer be used as a reason for stockpiling nuclear materials.

Conclusion

All of this work is intended to improve the human condition worldwide, as well as allowing us to co-exist with our planet in a less harmful way. Profits can be made from the development of this system in other ways. But the use and propagation of this system itself should be done with an eye toward philanthropy and human kindness.

Chapter 4 - Proactive Weather-Based Water Management

In order to take steps forward in these new efforts to treat these symptoms, let us take a brief look at what we have already done:

The scientific community and a large number of people are already aware of the potential devastating effects of climate change. Many of them are taking action in research and in urging for change.

Already there are many places around the world who keep sustainability in mind when constructing new buildings and cities.

- Beautiful rain gardens run through many towns, resulting in artificially created streams often lined with carefully thought out native plant choices.

- Many cities have created basins for retention and detention of storm water to help prevent flooding and erosion. Plant-filled bioswales are a better alternative to bare cement swales running along the sides of streets.

- Rooftop gardens adorn specially built or very sturdy buildings, increasing the amount of space available for oxygen giving plants, and for the natural cleansing and absorption of water.

- Individual enthusiasts create and use systems of rain barrels for non-potable and even drinkable water.

- There are even experimental communities in Sweden and Germany built for the purpose of improving resource management as a whole. They're beautiful, to boot!

All of these efforts and more make a difference when they are added together.

Now, in conjunction with these methods, it's time for us to take it all a step further. I've mentioned parts of this in previous chapters, but here is where I'll put it all together.

The Threat To Our Freshwater Resources

There are a number of different ways that we find and extract drinkable water. Once this freshwater has been taken from the ground or drained from a larger source, it is treated in water plants to be made safe for us to use.

- Underground wells provide water to areas such as small communities or individual homes.
  While these have provided drinkable water in the past, there is the increased threat that this water can become contaminated if that water table is tainted by soil salination, methane migration, and possibly by hydraulic fracturing (fracking).

- Creeks, rivers, lakes, and other natural resources of fresh water are filtered and treated for human use. These water sources replenish themselves through the water cycle.
  Unfortunately, human water usage is increasing due to our increased populations.

- Glaciers also provide us with a large amount of fresh water.
  But with the glaciers melting in the polar regions, this freshwater is mingling with the ocean water and becoming undrinkable. This also causes the sea levels to rise which contaminates some of our coastal sources of freshwater with that undrinkable seawater.

Our sources of freshwater are limited. It's also getting increasingly difficult for those sources to be replenished by the natural water cycle.

We will have to reduce our reliance upon those freshwater resources, and find alternatives. This will allow those freshwater resources to recover again.

The following picture has become my favorite demonstration of my reasoning behind developing new filtration technology and moving to the treatment and use of sea water.

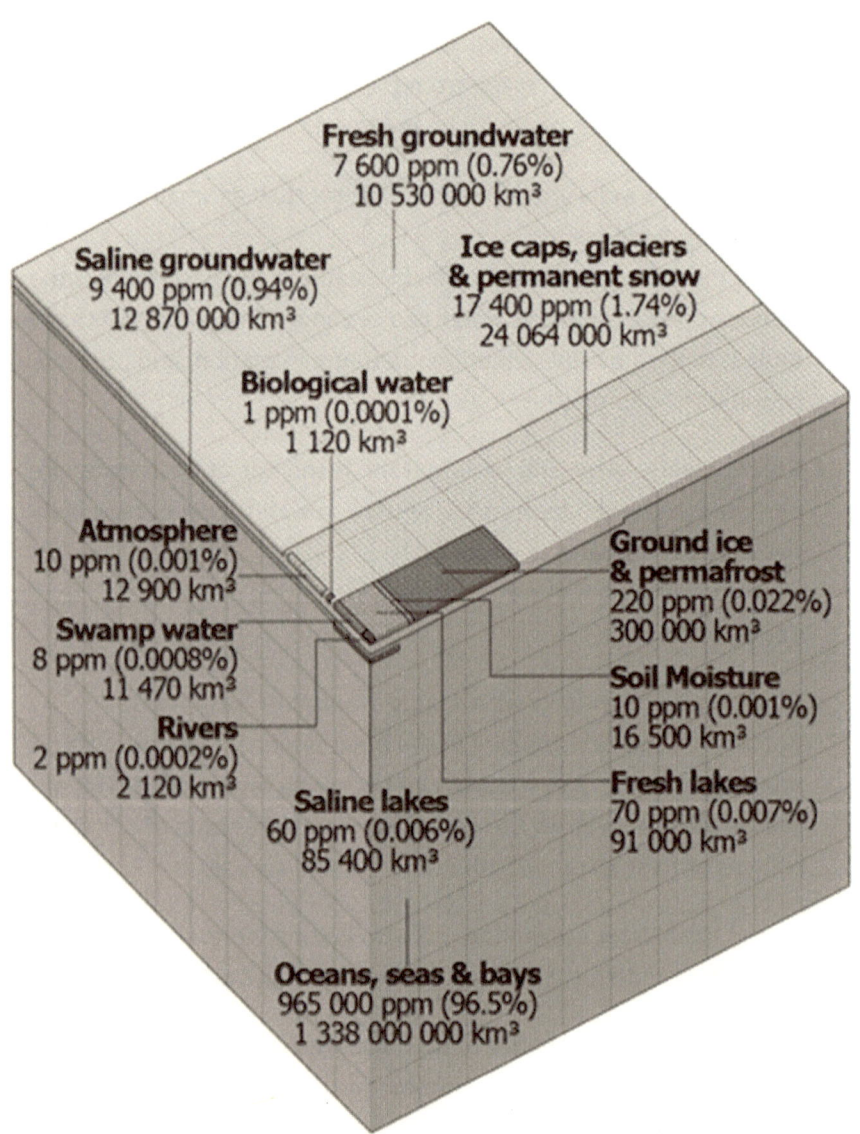

Fresh groundwater
7 600 ppm (0.76%)
10 530 000 km³

Saline groundwater
9 400 ppm (0.94%)
12 870 000 km³

Ice caps, glaciers
& permanent snow
17 400 ppm (1.74%)
24 064 000 km³

Biological water
1 ppm (0.0001%)
1 120 km³

Atmosphere
10 ppm (0.001%)
12 900 km³

Ground ice
& permafrost
220 ppm (0.022%)
300 000 km³

Swamp water
8 ppm (0.0008%)
11 470 km³

Soil Moisture
10 ppm (0.001%)
16 500 km³

Rivers
2 ppm (0.0002%)
2 120 km³

Fresh lakes
70 ppm (0.007%)
91 000 km³

Saline lakes
60 ppm (0.006%)
85 400 km³

Oceans, seas & bays
965 000 ppm (96.5%)
1 338 000 000 km³

The cube represents all of the water in the world. The thin slivers at the top are various types of water on Earth. They are mere fractions of a percentage for each water source, except for a few. Added together, the treatable fresh water totals less than 5% of the water on our planet.

The oceans, seas, and bays are the vast majority of water on the Earth. More than 95% of the Earth's water is in our oceans! This amount is only increasing as our largest source of freshwater, our glaciers, melt into the seas.

If we can let nature reclaim the far rarer natural freshwater for its own use in supporting wild life and the land we rely on, we'll run into fewer problems with floods and drought. When the soil is no longer overly dry, it will be able to absorb rain water properly without run-off, and our crops will grow the way they used to.

As for the oceans themselves, we would need to inject only a small percentage of its waters into this water system for all of humanity over our entire planet! The rest would be left undisturbed for marine life to enjoy.

And enjoy it they would. Marine life would have less chance of dying from our reliance on fossil fuels. There would be fewer accidental oil spills if we are able to rely on the energy coming from my suggested hydroelectric resources.

The Current State of Water Management

Current sewer and water pipe systems are used by water and waste management systems in every town or city. Each separate localized system of pipes uses water that was drained from local freshwater sources and treated by the water management plants.

As I discussed in previous chapters, this is a system on the micro scale. This scale moves from larger areas like reservoirs and the water management buildings, to smaller areas like individual homes, sprinkler systems, fire hydrants, water fountains in public parks, etc.

The new macro scale of water transportation that I am recommending takes this in the other direction. This is water transport between water management buildings and reservoirs, and transporting it to other areas of the country where it is needed more, or put into larger tanks for storage.

Here is a good example which shows how this system can work.

Some states like Florida have an extensive system of canals. These canals were made to handle the amount of water present in land so close to sea-level. The Water and Waste Water Services deliver water to residents, and oversee the canals that support aquifer recharge and flood management.

These canals have been a very large part of protecting Florida from flooding, and have relieved flooding conditions much faster than would have happened without them in place.

When Florida is threatened by a hurricane, water can be shunted from threatened areas to reduce the chances of flooding. This gives flood waters room to drain out of populated areas.

There's little that can be done against a 20-foot storm surge when you're living inches above sea level, but in Florida's case, every little bit helps once that initial storm surge passes. The various water management districts are enormously impressive! Take a look at the South Florida Water Management District for a fine example.

The Florida canals are a source of inspiration behind the pipes infrastructure in this system. They've demonstrated that the system can work. Now we need to take this a step further, building a more controllable yet less intrusive system.

Proactive Weather-Based Water Management

We need to update our scale of tracking the weather. This means Global weather analysis instead of only tracking localized systems.

Agencies like the National Hurricane Center and other groups already track the weather globally. And, on top of that, the World Meteorological Organization announced that new geostationary meteorological satellites will result in an enormous increase of pictures and data of Earth's weather.

Here are some examples of imagery taken from one of the satellites called Himawari. I never realized I was this much of a weather nerd until I looked at some of these amazing REAL-TIME images!

Our next step is to take all of that weather-based data and analyze it all with an eye toward the never-ending chain of cause and effect. Here is another instance where the proper programming and use of a supercomputer could result in enormously accurate predictions of our weather.

Many years ago, it was nearly impossible to accurately predict the weather. But if we apply mathematics and pair it with the new detailed data we have access to, we can take into account all variables, even ones from across the globe.

Supercomputers make this possible now. It's time to use this potential. Not only can they track the weather as I described above, but as I will detail below, they will be able to analyze whether other actions are necessary as well.

Soil Moisture Levels

On the surface, soil seems to be uncomplicated (pun intended). However, soil science is very complex. As complex as the enormous number of variations in the different materials that soil is made out of, and the living things occupying it.

All of those variables have potential effects on how well soil can absorb water, and how much water that soil can retain before it becomes too saturated to absorb any more.

There is also the factor of how dry soil can become before plant life can no longer extract water from the ground. These plants of course die, leaving the nutritious top soil with little protection from the baking sun and harsh winds.

This type of drought results in other issues like dust clouds, or soil baked so hard that water runs off of the top of the surface instead of absorbing into the soil. When a drought-ridden area finally gets rain, one would expect the ground to drink up that rain - problem solved, right? Instead, in some of these areas the water runs off and gathers in low lying land, creating flash floods in an area that is still in a drought!

Smart Irrigation

Smart irrigation and soil aeration can help to maintain soil moisture levels. These are two processes that have been in use for a very long time. However, as with most other things, it is about time for us to take our improvements in technology and information, and re-evaluate how to apply these processes and more to improve our relationship with our land.

Smart irrigation refers to using measurements of soil moisture and other factors when scheduling irrigation to areas like farmland. Other factors would include temperature and humidity levels; it makes little sense to water crops during the warmest part of the day when a lot of that water can be lost to evaporation.

Already we used improved technology to be able to measure the amount of water present in soil, either from nearby, or even using the newest satellite technology presently available to measure much larger areas of wild forests.

Soil Aeration

Soil aeration is the process of using equipment to either puncture soil using spikes (spike aeration), or the removal of a couple of inches of cores of soil (core aeration). Both of these methods are vital ways to improve drainage.

After aeration, drought-dry soil can absorb more water without forming puddles or becoming a field of mud. Even healthy land can benefit from proper aeration, improving water and nutrient infiltration and allowing roots to grow deeper.

I'd advise increasing the use of these methods as a public service, especially in areas undergoing drought. By aerating dry soil before irrigation or rain, there will be far less run off into flash flooding, and a far more efficient way of ending the drought.

It isn't only farms that would benefit. Soil aeration can be done much in the same way that the grass on public land like highways and parks is mowed; by using machines. Aerating the soil properly will mean less flooding of the roads. If aeration machines or services were made available to homeowners, it is entirely possible that there will be far less residential flooding as well.

By using smart irrigation after aerating the soil, we can take water delivered from the Water Houses, and keep the soil saturation at a carefully measured level. This process will support plants while allowing that soil the flexibility to be able to absorb more water when strong storms pass overhead.

Conclusion

In this chapter, I discussed our current methods of water management, and how we can move to the ideal

The mathematician and meteorologist Edward Lorenz was credited with a well-known description of how a butterfly, flapping its wings in one part of the world, could affect the weather on the other side of the globe.

With the new tools we have at hand, we are getting closer to being able to solve this mystery ourselves. In the far future, we may even be able to apply it to certain levels of weather control, defusing Category 5 storms into something more bearable, or perhaps even steering normal rainfall into places where it is much needed.

Chapter 5 - Preventing and Fighting Wildfires

Firefighting in general can benefit from the use of this system. While I'm glad to be able to say that I've rarely heard of any problems with fire hydrants in the mainstream news recently, it doesn't mean that there haven't been any.

Also, any steps to improve the water infrastructure and prevent future problems are good steps to take. Any and all possible action should be a priority when it comes to protecting our emergency personnel and our populated areas.

The cycle of life in nature includes forest fires as a normal thing. When trees and even animal life burns and dies, the resultant materials provide nutrients to the forest's soil. But the recent increase in UV radiation from holes in the ozone layer, the increases in temperature, and the drought in many areas are causing more forest fires than there have been even a few years ago. We need to bring this under better control, and in a way that makes out-of-control fires less likely in future.

Current Methods

While our current methods are effective, they could use some improvement. Especially with the increased resources that the system that I'm suggesting would provide.

Here is a view into just some of the current conditions and hard work.

Fighting Wildfires

- Brave volunteers fight fires from the air and on the ground. Some die or are wounded. According to the Bureau of Labor Statistics, their pay scale varies widely, depending mostly on location.

- The locations for dousing the wildfires are strategically determined to best effect for containing a blaze. This is done by limiting where the fire can travel, then letting it burn itself out by running out of fuel like trees, plants, and trapped animals. This is done using several methods like Fire Lines or Controlled Burns.

- Chemical retardants are often used by aircraft flying over a wildfire. These chemicals have been created and tested for their efficiency in making an area unable to combust using different methods. The material can come optionally colored so that it's easy to see what areas have been treated by other aircraft. Some of the colored retardants will fade over time in sunlight.
An example of a commonly used fire retardant is Phos-Chek. From their website, you can see that there advantages and disadvantages to using chemical fire retardants in wildfires. The majority of the retardant is water. Other materials include thickening agents so that the water doesn't disperse, and certain fire retardant chemicals that can biodegrade or be used by plants as fertilizer. Unfortunately, toxic levels of the chemicals can poorly affect the environment.

- Portable tanks of water are transported closer to fires for use by the ground firefighters. Helicopters use water from any nearby water sources like rivers and streams to dump water in strategic locations.

- Firefighters on the ground comb the forest to make sure there aren't any embers or hot spots that can re-light a fire.

Preventing Wildfires and Aftercare

Many worldwide organizations are doing research on rehabilitating forests and better ways to fight fires on the ground and in the air.

- **Already there are early-warning fire information systems being developed.** This research should be encouraged and assisted. For example, the AFIS real-time system is reportedly very successful in detecting wildfires. By visiting their website, you can access their AFIS detection system and see the location and severity of wildfires worldwide!
America has one too, run by the USDA/U.S. Forest Service.

- **Smaller but important measures are being taken.** Organizations such as the U.S. Forest Service is encouraging homeowners near large forests to make sure that there are 100 ft. of space cleared between their homes and the edge of the forest.

The homes themselves should also be made of fire-resistant materials and design. An example is this man who survived a forest fire in his fire-resistant home.

- **After a fire has been put out, the job isn't finished**. Additional work is done to prevent soil erosion and water run-off due to the defoliation and other damaged caused by the wildfire. This includes tasks like building water bars to redirect any water run-off onto a preferable path, avoiding trails and roads, and planting new trees.

Improvements Using The Proactive System

I have a few other suggestions for improvements that would take separate research from my system. With the increase in the number and intensity of wildfires the past few years, the added improvements are necessary if we are to protect our people, private property, and the natural resources found in our remaining forests.

Protection of human life is the first priority as it should be. But with leaps in improvements to our current methods, we can ensure human safety. Once that is done, we can turn our priority to extending greater levels of protection to the forest and wildlife as well.

Fighting Wildfires

- **Stored water from special lightweight tanks can be on hand for nearby fires**. We can transfer water from our storage tanks to portable tanks. These tanks can be transported by air to be used by ground troops, or the water can be loaded onto air vehicles to be dumped strategically onto the fire. Until a completely safe and organic retardant can be created, localized filtered/desalinated water is preferable for firefighting.

- **Invent/improve the fire retardants used in firefighting**. Make certain they have no lingering ill effects on plants and animals even at 'toxic' levels, and are safer and easier for human use. Suggested factors of additional improvement: Make the materials lighter and more concentrated so that more can be carried in the same amount of space as the mixtures today. Make a retardant which expands upon being heated up or expands in a controllable chain reaction when hit with water so that more of the area is covered.

- **Create flexible and fireproof temporary pipes**. These can be extended or contracted as needed, and stretched from linked water tank systems that are brought to the site of the fire. These pipes can optionally connect to any nearby pipes from our pipe infrastructure, and use water from the system to fight nearby fires. Temperature control would come in handy here so that water can cool overheated areas as well as douse them, and also provide drinkable water to the fire fighters.

- **Build temporary enclosed wind-breaks out of light, flexible, waterproof and fireproof material.** Smaller areas of fire can be carefully cleared of humans, and a wind break put up or possibly lowered by helicopters. That area can be doused without fear of the wind spreading the fire before before moving the windbreak to another area. Also, the wind break area can have a "roof" of the same material set on top of it, cutting off oxygen to the fire and helping to snuff it out that much faster, especially when paired with flame retardants that give off water.

- **Larger wind breaking walls can be put up** to help prevent large fires from gaining unexpected strength and jumping long distances due to a shift in the wind. Using the proactive weather tracking system, firefighters would be more informed as to when strong wind is likely to blow, and what direction it would head. The large wind breaks could be put up strategically, and moved in sections by helicopters as needed once an area has been put out.

- **Track drones interfering with air firefighting vehicles.** Drones need to be controlled. Some interfere with firefighting and rescue attempts, and because they are endangering normal air traffic. They can also be used very easily for violent purposes.
People need drivers licenses and they are required to be tested in order to drive on our roads. People should be required to register and follow

similar laws in the air for the safety of everyone.

Preventing Wildfires

- **We should minimize building pipes and tanks in forested areas** so that we don't interfere with wildlife and the natural growth of forests. The pipe infrastructure can be built along the borders of towns and forests. Areas with the greatest risk of fire could be considered for a pipe running through it, but the next measures I describe should help to avoid that necessity.

- **Satellite systems would measure important data** such as soil dryness, heat, humidity, UV levels, and holes in the Ozone Layer. (More about the Ozone Layer to come in future book!)
  This information can be used together with weather forecasts to determine if an area of forest should be treated to prevent wildfires from breaking out. We already have systems like these, but the increased amount of information and accuracy will be invaluable. Along Track Scanning Radiometers aboard weather satellites are able to detect wildfires, which show up at night as pixels with a greater temperature than 308K (95°F).[12] The Moderate-Resolution Imaging Spectroradiometer aboard the Terra satellite can detect thermal hot spots associated with wildfires, volcanoes, and industrial hot spots

- **Driest areas of forest that are at high risk of fire should be unobtrusively aerated and irrigated**. It'll take a longer time to do this without frightening wildlife, but it's necessary in order to preserve those creatures. Aerating the forest ground will also improve tree growth and lessen water runoff.
  As I mentioned in the previous chapter, soil aeration and irrigation can be done in populated areas as well, so ensure that they are being well maintained so that people are safe.

Optional Use of an Emergency Water System

Coastal Water Houses and a second system of pipes can optionally gather extra water from the oceans for use in fighting local fires or being used as an emergency storm levee. Special water tanks could be anchored into place prior to a hurricane or typhoon. By locating them on strategic areas of coastline, and filling them with water to further weigh them down, they would help protect against storm surge and coastal erosion.

Because the water is less filtered than water that traveled through the system of pipes, the water should be kept to the local area so as to not unbalance the ecosystem in other more distant areas. For example, the local sea water off of the coast of California could be treated and used to douse wildfires in the coastal areas of that state.

Of course, it is vital that any emergency water for treating fires must go through a desalination and filtration process to remove salt, oil, and other immediately harmful chemicals. Otherwise we would be doing what is called "salting the earth". This would make our land unsuitable for anything to grow. This, of course, would be devastating.

This additional minimally-filtered water system is best suited for dire emergencies only. It would assist in a minor way to avert the growing threat of ocean level rise, combat major life-threatening forest fires, protect against storm surge, and preserve property and wildlife areas.

However, if we consider this option, as with the normal pipes for gathering ocean water, we would need to make sure that the pipes do not damage our coastlines. This includes any coral reefs and other natural resources, interaction with other pipelines, or harming the surrounding wildlife while the emergency pipes are functioning.

Conclusion

In this chapter, I discussed our current methods of firefighting, and ways that the system I'm discussing in this book can be used to make it easier for our brave emergency personnel.

I'm glad to say that research is already underway on using satellites in order to measure soil dryness, and the human influence on deforestation. This should be encouraged, and resources from this system used with it in order to bring wildfires back under control.

Chapter 6 - Restoring the Earth's Glaciers

Here is where we can get really creative in using our new water pipes infrastructure to solve problems that are less obvious.

Removing water from places where there is too much of it is obvious. Transporting water to places that has too little of it is also obvious. Once we can do these things in a way that won't impact the Earth in a bad way, we can take the next important step.

What is the next step?

Consider the problem of our continual loss of freshwater resources. Glacial ice is the largest reservoir of freshwater on Earth. With glaciers dwindling and melting into the surrounding seawater, we are losing our valuable potable water resources.

Glaciers are a vital part of the Water Cycle on Earth. This is the method that the Earth uses to constantly renew and deliver freshwater to the natural resources like underground water, lakes and rivers.

When disruptions happen to the water cycle and there is less rain and snow, there is less water available during the spring melt, resulting in drought.

Alternatively, there are some areas where glaciers melt and flow into the rivers of an area, causing floods. Once these floods have passed, there is less future water resources unless the glaciers are replenished properly.

NASA is currently doing a fantastic job on measuring the current gains and losses in glacier mass, temperature changes, sea-level rise, and even charting the effects on arctic lifeforms! This information will be very important for our research in order to restore our glaciers.

The Water Cycle

The Water Cycle is a full circle, moving from water in liquid form to being evaporated into the air and cooled into rain or snow, the water being stored as snow and ice, and then melting back into water when the weather warms.

Glaciers are the step in the water cycle where water is gathered and stored during the winter. In recent years, there have been alterations to the behavior of the Arctic weather during the winter. In some areas of the world there has been far less snow than normal, while in others there are sudden surges of dangerously cold weather and storms where it had been more stable before.

Arctic Temperatures Moving Further South

These new shifts in arctic cold comes from a shift in warmer temperatures. This in turn can cause shifts in wind, and changes to the water and cold that emanates from an area.

Have you ever seen a piece of ice melting? If a heat source is located to the side, more melted water will pool on that side. If you put your fingers near the ice cube, a chill in the air will be easier to detect from the side that is being heated as the cold from the ice interacts with the warmer air.

The strange Arctic chills we've been experiencing especially over the last few years over winter are likely another symptom of the glaciers melting. Watching ice melt through heat sensors can show you a good example of this effect.

An entirely different aspect is how wind and a layer of cold glacier melt is reportedly allowing warmer waters to reach beneath some of the glaciers causing even further ice melt below the surface!

This is why a careful study must be done with our improvements in computer and information technology. We need to add the pattern of glacier advancing, retreating, glacier melt, and the interactions between cold and warm air and water into our Proactive Weather Management system. Without continual analysis, lesser known areas such as the poles have a good chance of throwing off the equations.

What are the Real Problems with Melting Glaciers?

As I mentioned earlier, ice caps and glaciers are a large part of the Earth's freshwater supply. However, as they melt and that freshwater goes into the ocean, it causes two major problems on top of all of the others I just mentioned:

1. We lose that amount of freshwater when the glacier melts into the ocean and mixes with sea water. The majority of water on the Earth is sea water rather than freshwater which can be more easily filtered and used by humans and wildlife.

2. The melted glaciers cause sea-level rise. This causes the sea water to seep into the freshwater table further inland, reducing the amount of freshwater available for use even more.

Fortunately, this system I've described in this book will give us a chance to restore the balance ourselves.

Using the Water Houses and Pipe Infrastructure System to Restore our Glaciers

With careful research on procedure and the type of water needed, we can use the water from areas where glaciers have melted, filter it, and replenish those icebergs at the poles and other locations. However, we can't build permanent Water Houses or even pipe infrastructures at the North Pole because there is little solid "land" there. Water tanks and flexible pipes would need to be used instead of the more permanent Water Houses and pipe infrastructures. We could build more permanent structures at the South Pole as there is actual land there, but I'd encourage that to be minimal so as to not interfere with the wildlife.

With strategic maneuvering, ice caps on mountains can also be replenished but that would need to be done with great care so as to not prompt an avalanche! Beijing plans on creating man-made snow for the 2022 Winter Olympics. I'm sure they'll be plotting the work out carefully, but I hope they'll be sticking to natural man-made snow rather than adding artificial agents (more below).

Some ski resorts have used "snow guns" or "snow cannons" for many years in order to supplement their natural snow. This involves a more complex process than simply having water in the air freezing into snowflakes at the right temperature. Some Snow cannons mix water with a nucleating agent to help freeze the water more quickly.

Obviously, if we are to use part of this process to help restore the glaciers, there can be no additives to the water. Fortunately, by doing this in the coldest parts of the Arctic zones, there will be little trouble getting water humidity in the air to freeze into snow without chemical assistance.

Water Houses on the nearest part of the continents can oversee the process so that we can strategically restore areas that need it the most, and are the safest to do so. Adding snow weight to a weakened ice sheet may result in that ice sheet cracking and falling into the ocean, for example.

The icebergs that we replenish will help to lower the temperature of the oceans and moderate the hurricanes spawned by higher water temperatures.

The lowered water temperatures will also help stabilize ocean life which is currently undergoing migration north to get out of the hotter waters near the equator.

Tips on How to Start Glacier Restoration

Glaciers form where the accumulation of snow and ice exceeds the amount of water lost due to melting into the sea, evaporation, or pieces of ice chipped from glaciers due to wind or sudden collapse. This water loss is the measurement that we must overcome. Simply, we just need to add more snow and ice to the glaciers than the amount of snow and ice that is lost.

The best time to restore glaciers is when we have the assistance of lower temperatures during the winter. This also needs to be done during a period of time in the coldest parts of the arctic where there is still sunlight so that it's easier for humans to see and to survive in those areas, especially in the North

Pole where we have less land stability. Before winter would be best so that the winter cold would set the glaciers after our visit.

The further toward the Earth's poles that you go, the less sunlight there is during the winter months while the tilting of the Earth's axis directs the poles away from the sun. Some areas of Earth see absolutely no sunlight for months. Fortunately for us, we can concentrate our efforts on the glaciers at one pole at a time during this window of opportunity.

Research and strategic decisions will allow us to best decide which areas and what types of glaciers to replenish. Different types of glaciers have different functions in the arctic regions. Different areas may have much stronger winds-too strong for us to take advantage of. Instead of just freezing water into snow in the air and letting it fall, strong arctic wind storms could scatter the snow randomly.

We could get creative and put up specially made arctic wind blocks like the fireproof ones I suggested in the chapter about preventing and fighting wildfires.

Restoring filtered water to the glaciers will ensure that the water cycle can continue with our help. The water cycle will replenish the natural sources of freshwater, while humans become more able to depend on filtered sea water in a mostly closed system.

Since glaciers are so vital to the entire planet and affect us all, this is really more of a global project than something that is done by one country.

Conclusion

Restoring our glaciers will help to treat many symptoms of climate change:

- Glaciers help to moderate the world's temperature in the air and oceans.

- Glaciers are a source of freshwater that flows over and under the land to natural water sources every spring.

- Reclaiming the glacier water that has fallen into the oceans and melted, and carefully desalinating and filtering that water back into the glaciers - especially glaciers over land - will prevent much of the sea-level rise we're currently threatened by.

- By preventing sea-level rise, we lessen the threat of sea water seeping into our freshwater tables underground, and preserving the coastlines and freshwater coral reefs.

- By restoring clean water to the Earth's water cycle, we are helping to prevent flooding and drought by bringing the amount of water contained in soil back in balance. The soil will be able to support crops, and there will be fewer flash-floods due to water run-off.

With this information, you can see that it isn't only the problem of glaciers raising the level of sea water over a long period of time. There are a lot of additional problems that can be solved if we protect and restore our glaciers.

Summary and Conclusion

As we discussed, in recent years flood and drought has seemed like a senseless and overwhelming problem. Fortunately, it has been coming into the public's awareness now more than ever. More people are searching for information on the web in the hopes of figuring out what can be done about it.

In this eBook I have discussed a system which can be the answer that we need. There are a number of parts to the system, and there are a number of actions that we need to take. Here is a summary of some of those ideas.

Water filtration

Currently our methods of water filtration are thorough enough to make water safe to drink. However, for the system we need to take water filtration a large step further. In order to move water between areas, we need to have water pure enough that there won't be any differences in chemical composition or microorganisms that can throw off two different ecologies in two different areas of the country.

There would be at least three stages of filtration. The first where water is taken in, either at the oceans, rivers prone to flooding, or storage tanks that have been used for emergency water drainage during a flood.

The second stage is in the pipes infrastructure. A large series of filters along the entire length of the system will remove the majority of impurities.

The third is in the water houses where the last of the waste material will be removed.

Not only does water have to be cleaned, but it also needs to be desalinated. If we are to preserve and replenish the freshwater resources, then we need to start using water from the biggest resource on the planet. The oceans.

We have started using or polluting a majority of the freshwater available. But freshwater is only a small fraction of the water on the planet. Our using water from the ocean for our survival should not impact the ocean greatly- as long as we are careful about how we do it!

Pipes Infrastructure

We already have a system of pipes on the micro scale, delivering water from water management plants to individual buildings and homes. We also have a pipe infrastructure to transport things like natural gas and even oil across the country.

As I discussed, we need to build a system of water pipes on the macro scale. These pipes will serve many functions even beyond transporting water from areas of threatened floods to areas that are suffering drought.

- **Water tanks** will be part of the system, storing extra water, or remaining empty to be filled if the water pressure in the pipes gets to a level that needs to be reduced. Emergency tanks can be used to quickly drain water from an area undergoing an unexpected flash flood.

- **The pipes** themselves will have systems inside them so that information can be sent to and from Water Houses. Water can be used at any point in the pipes to help fight wildfires. To do that there needs to be filters added to the pipes themselves.

- **Water filters** would be placed periodically along the entire pipeline. As the water is transported, doing so will act as the second stage filters mentioned above.

- **Sensors** also would be a part of the system in order to maintain proper water and air pressure inside the pipes infrastructure. The contents of the water can also be monitored in the pipes, alerting us of any toxins in the water either through accident or through deliberate acts. We would also be notified if there was any damage to the pipes either by nature or tampering.

- The same **taps used to access water** in order to fight wildfires could also be the valves use to drain water from the system that has become toxic. Other physical structures that would be part of the pipes infrastructure would include **pumps** to act like a signal repeater to keep the water moving. **Controllable valves** are placed so that water can be directed as needed away from flood threatened areas and toward areas undergoing water shortages or threatened by wildfires.

- The pipes would need a form of **temperature control**, either through devices like the water filters or through the material of the pipes themselves. This would prevent freezing or evaporation of the water. This temperature control can also be used on the micro level to help with snow removal.

The entire system can be **self-powered with hydroelectric energy generators** placed periodically along the pipes infrastructure. If the generators are powerful or numerous enough, the water could generate enough power to slow or halt our dependence on fossil and nuclear fuels.

This way, the very act of transporting the water as we need it will also filter the water, test it, and generate energy.

Water Houses

Water Houses would be operated by trained specialists cleared by security. These people would be in charge of overseeing the computers and processing the results of all of the information coming in from the entire system. They would also monitor the weather reports and proactively prompt the system to move water as needed, away from areas of threatened high concentration and toward areas of low concentration or storage.

Water Houses can also have their own set of pipes for energy generation and localized water filtering. This would enable the Water Houses to be off of the grid independent of local power.

Water Houses also need to be secure from hacking, and need to be in constant contact with each other. I suggest using their own separate intranet rather than being connected to our world wide system. However, security specialists would be able to better make that decision.

Proactive Weather-Based Water Management

Our sources of fresh water are being threatened from many sources of pollution, and from nature's reaction to changes in temperature which is prompting sea-level rise.

If we learn to monitor the weather as a single global system rather than only watch localized systems, we would be able to predict changes in weather with far greater accuracy. We have supercomputers capable of this already.

This monitoring would include the paths and strength of storms. As a result, we would be able to manage our water resources, moving water out of an area that is about to experience a strong rain storm, for example.

If we begin to shift our dependence more toward ocean water, it would allow our freshwater resources to naturally replenish themselves as they are used less.

If we allowed wildlife sole use of the freshwater, the land and the water cycle would be able to start recovering from our recent overuse.

If we manage our filtered water resources in a way that assists in the land's recovery through methods like smart irrigation and soil aeration, humanity won't be at the mercy of flooding, drought, weather and wildfires.

Preventing and Fighting Wildfires

We currently have so many brave people worldwide whose job it is to see to our safety. Some of those people fight fires in populated areas, and over thousands of acres of forests. They do a fantastic job with our current equipment and methodology! But with this system and other innovations, their job can be made safer and easier.

Water from nearby storage tanks, Water Houses, and pipes bordering the forest can be used to help fight forest fires. Additional tools can be created, such as flexible and fireproof temporary pipes in order to bring that water where it is needed more quickly. We can also stop wind from strengthening fires and helping them to travel by erecting fire and waterproof wind breaks. Current fire retardant chemicals can always be improved on in ways that benefit wildlife.

The threat of wildfires can also be far more accurately predicted with the use of computer technology predicting storms and lightning strikes. Satellites can scan the Earth's surface and advise about the areas of forest that are the driest. These areas can be treated with aeration and smart irrigation so there is less chance of a fire outbreak.

Restoring the Earth's Glaciers

As most people have no doubt heard, a number of well charted glaciers at the North Pole and the surrounding continents have been losing mass which hasn't been getting replaced by the precipitation part of the water cycle. The South Pole is colder and less humid, but is still being affected by changes in the weather.

Water from sea and land glaciers is being added to the oceans. This both reduces our supplies of freshwater, and contributes to sea-level rise.

While sea glaciers are partly already in the oceans' mass, **the glaciers on land are not a part of the ocean at all!** When they melt, those glaciers have been the largest contributers to sea-level rise.

This sea-level rise also causes sea water to seep into some of our freshwater reservoirs across the world, reducing our freshwater supplies even more!

After carefully analyzing the effects on the wind and the glaciers as they are now, we could encourage snowfall without seeding the clouds with chemicals. In the same way some ski resorts create their own snow, we could spray desalinated and filtered water taken from the ocean into the air and allow the coldest temperatures to freeze it into snow. This would give the glaciers material to be replenished.

Future Jobs

While there might be a lot of detail within this book, the proactive system will require research and invention in order to make this become reality. Every aspect can be improved by the scientific community as a whole, together with construction and engineering corporations, the computer industry, utilities, not to mention those who can approve these changes to begin with. The expertise and determination of millions of people is a valuable and much needed resource.

While pipelines are known for being able to be run with minimal staffing, this proactive system as a whole will need more people, even after it is built.

Specialists will be needed to maintain the water quality, monitor water pressure, direct water to where it is needed, upkeep the physical condition of the system, monitor security, and regulate and distribute the energy generated.

In addition, there will be people required to investigate integrating these systems into other countries who request it once we have proven its success in our own.

Building in Developed vs Less Developed Countries

There is the option of creating this proactive system in a country that is in dire need of clean water and green power among other things, if they are given the choice and request it.

If there is no system currently in place for filtering and transporting water, and no access to stable power generation, then it would actually be easier to create this proactive system. The jobs created would bolster their country's economy.

Most importantly, the prospect of saving people's lives rather than fighting for resources can bring the world closer to working together peacefully as well! It may also encourage migrants to return to their homes if living conditions improve and fighting dies down.

In the US, we will be replacing and building on top of an already existing system. This will give us shortcuts in some areas, but it will also mean extensive construction in other highly populated areas.

At the same time, care will need to be taken in order to preserve the natural ecologies and unspoiled land on our planet. There is less of it every year.

This proactive system shouldn't require cutting through those areas. The simple action of water traveling through pipes is an advantage rather than a problem. The trip can be made longer in order to avoid wildlife and use the added distance to filter water and generate more hydroelectric energy!

Treaties for our Future World

If the system is made well, it could make arid countries more livable for those that wish it. There would be more food and water for everyone.

For landlocked countries, international treaties would need to be established in order to implement this system. Pipes would need to service those landlocked countries after traveling through the coastal ones. This would hopefully result in leaders finding it easier to continue working peacefully with the other countries in the area.

War-torn countries could also be made far more suitable to live in, if only their rulers could come to peaceful agreements. By making treaties and building this system in their countries they could improve the quality of life for their people. An example would be Maslow's Heirarchy of Needs. By helping to deal with these needs, the proactive system might also help to reduce the strain of the current migrant crisis if we act quickly, or to avert future ones.

Hopefully with these blocks to survival lifted, humans would be happier worldwide and look more toward improving themselves instead of turning to violence.

Research Opportunities

For those who are interested, here is a list of the opportunities for research that I mentioned in my ebook. All of these inventions could be used to improve our world, and the human condition.

- A water intake system crafted to have minimal impact on the oceans and rivers and other areas that deliver water into the system

- Water filtration modules capable of cleaning water of all impurities

- Water Waste reclamation methods to re-use or neutralize waste instead of needing to find somewhere else to put it.

- Improve methods to assist with water retention in soil such as aeration and proactive irrigation

- New pipes materials. Strong, possibly flexible, assisting with temperature control, able to be constructed as modules to add filters, sensors, valves, energy generators in the future as the structure grows or is modified.

- Real-time sensor analysis of water content, pipe and air pressure, energy generator output, temperature.

- Security system to report direction of water in the pipes, any physical issues with parts of the system, etc.

- Hydroelectric generators that can be periodically installed along the pipes. Each generator would help to power the system. Efficient hydroelectric generators can become a power source for the world.

- A new global weather tracking system that takes all weather into account. Weather could be accurately predicted days or weeks ahead.

- Fireproof and waterproof temporary wind breaks that can be secured in the ground by helicopters to help stifle fires and stop problems from high winds feeding the flames.

- A foolproof method/machines to take cleaned water and strategically make it into arctic precipitation to restore mass to our glaciers.

Hope for our Future

While this system of pipes and Water Houses can be a fantastic tool for evening out water usage in a single country, this would be best as a world-wide project. As I mentioned, there are some land-locked countries that CANNOT be held hostage to paying too much for the right to filtered water from our world's oceans. Everyone can contribute to the effort, and by doing so, given access to the water that every human on Earth needs to survive.

The filtered water can be delivered to storage tanks and protected reservoirs. Water instead of oil can be transported on trains to hard-to-reach areas. It can be used in farms, supplement water that our population lives on, and if an innovative technology is developed, it can be used to create energy.

By evening out the level of potable water on Earth, we should see a sharp reduction in such hardship as dying crops and livestock.

- If our water filtration system is effective enough, there should be nowhere that would suffer a drought.

- If our water transportation system is responsive enough, we shouldn't hear news of a flood destroying people's lives.

- No towns would need to rely on bottled water while contaminated water comes out of their taps.

- No populations would be without water or power after a hurricane passes overhead.

- No life or property would be needlessly lost to waters rising out of control.

- As this project develops and is reported in the news, there will be a greater climate awareness encouraging us all to work together.

If the system is made well enough that it can be used in other countries, we may well see a time in the future when no humans are thirsty.

Ever again.

Amanda Rothman's Resources

If you want to be notified when my next books come out, or if you just want to contact me or learn more about what I'm doing, please take a moment to visit my website:

www.proactive-action.com

I'd like to invite you to sign up for my newsletter. You'll receive updates about contests, my future books and opportunities to get copies of them for free when they are released. You could even be the first to see previews of my books, and give feedback, becoming an influence on these global spanning projects!

Joining in and supporting these ideas would also make it clearer to the government how much approval there is, increasing the chances that they are put into practice!

I would love to hear from you if you have questions or ideas involving the system I have written about in my book. I make it a practice to reply to each message that I receive.

If you'd care to leave a review for my book on Amazon, it will go a long way toward helping to get the word out about my ideas!

Here is a direct link to my Amazon review page: http://amzn.to/1P3Jjir

Amazon Author Central

You can find my books listed in my Amazon Author Central page here:

http://amazon.com/author/amanda_rothman

Writing Your Own Book!

If you're interested in learning the process that helped me to write this book, learn more about the Self-Publishing School by watching the free videos here!

If you join the school through my affiliate link, you'll be doing two things:

1) You'll start your journey to writing a book you've always wanted to but didn't have the time to do (even if you aren't sure how to go about doing it).

2) You'll be further supporting my being able to write these books to help treat the symptoms of Climate Change, and to help encourage others to come up with ideas and to take action!

If you're interested in a helpful review of the Self-Publishing School, and what the course includes, I have posted information here.

Link Appendix

Introduction:

http://ca.water.usgs.gov/projects/central-valley/delta-mendota-canal-subsidence.html

http://www.cop21.gouv.fr/en

http://w2.vatican.va/content/francesco/en/encyclicals/documents/papa-francesco_20150524_enciclica-laudato-si.html

Chapter 1

http://water.epa.gov/action/importanceofwater/upload/Importance-of-Water-Synthesis-Report.pdf

http://www.npr.org/2015/07/30/427839942/ap-study-finds-viruses-linked-to-raw-sewage-in-rio-de-janeiro-olympic-waters

http://water.epa.gov/drink/local/index.cfm

http://water.usgs.gov/edu/drinkseawater.html

http://www3.epa.gov/chp/documents/wwtf_opportunities.pdf

Chapter 2

http://www.pasadenastarnews.com/environment-and-nature/20141021/leaky-pipes-waste-millions-of-gallons-of-water-in-la-pasadena-and-san-gabriel-valley/1

Chapter 3

http://waterwatch.usgs.gov/wqwatch/

Chapter 4

http://sustwatermgmt.wikia.com/wiki/Sustainable_Water_Management_in_Action:_Project_Examples_from_the_U.S._and_Abroad

Picture credit: https://en.wikipedia.org/wiki/Water_resource_management

http://www.sfwmd.gov/portal/page/portal/xweb drought_and_flood/canal and structure operations

http://www.nhc.noaa.gov/

https://www.wmo.int/media/content/wmo-welcomes-new-generation-meteorological-satellites

http://ds.data.jma.go.jp/mscweb/data/himawari/

Chapter 5

http://phoschek.com/product-class/fire-retardant-for-wildland/

http://www.afis.co.za/

http://activefiremaps.fs.fed.us/index.php

http://www.firewise.org/wildfire-preparedness/be-firewise/home-and-landscape.aspx?sso=0

http://www.npr.org/2015/08/26/434821436/firefighters-get-the-upper-hand-in-washington-state-wildfire

http://www.spc.noaa.gov/

http://dx.doi.org/10.1080/2150704X.2013.862601

https://en.wikipedia.org/wiki/AATSR

https://en.wikipedia.org/wiki/Moderate-Resolution_Imaging_Spectroradiometer

https://en.wikipedia.org/wiki/Satellite_temperature_measurements

Chapter 6

http://water.usgs.gov/edu/watercyclesummary.html

http://www.nasa.gov/content/water-and-ice/

Summary and Conclusion

https://en.wikipedia.org/wiki/Maslow%27s_hierarchy_of_needs

Amanda Rothman's Resources

Amazon Review page for this book: http://amzn.to/1P3Jjir

Amazon Author Central page: http://amazon.com/author/amanda_rothman

Self-Publishing School videos and program:
https://xe172.isrefer.com/go/SPS/athenakt/JoinSPS

Review of the Self-Publishing School: http://blackbootdiaries.com/self-publishing-school-a-brief-history-of-my-time-and-a-review/

www.ingramcontent.com/pod-product-compliance
Lightning Source LLC
Chambersburg PA
CBHW021439170526
45164CB00001B/309